BR ~ SULZER
CLASS 44 · 45 · 46
DIESEL-ELECTRIC LOCOMOTIVES

by
Julian A. Walker

CW01455839

Class 45s in the final stages of construction at Crewe works in grey workshop primer finish. These are from the batch D68 to D107. [J. R. Carter]

© copyright 1976   ISBN 0 85153 2934

printed in Great Britain by H. E. Warne Ltd, London and St. Austell

for the joint publishers the **DIESEL & ELECTRIC GROUP** and

D. BRADFORD BARTON LTD . Trethellan House . Truro . Cornwall . England

INTRODUCTION

Of the many main line locomotives ordered in 1955 and later in 1957 by British Railways, the 'Peaks' stand out as one of the most interesting in their development and design as well as in their good service record. Heavy and powerful, they play an important role on BR. Although this book is an attempt to give a detailed view of the locomotive's construction and operation, it is not meant to be a complete picture as the story has not yet finished. After some fifteen years' service, the 'Peaks' continue to do well and to work hard and it would be impossible to try and set everything down at this stage of their lives. Yet this volume fills a gap for those who want to have a detailed, yet practical, understanding of this remarkable series of locomotives, and know the type of performances that can be expected from them.

Several people have been of tremendous assistance in the preparation of this book, in particular Mr. A. Tayler, Manager, Diesel Traction Division, Sulzer Bros. (London) Ltd., without whose help this book would not have been possible. I am also indebted to my fellow members of the Diesel and Electric Group, particularly John Howie and M. J. Oakley, who has very kindly made available some of his best train logs and valuable observations on the performance of the 'Peaks'. I would also like to thank Mr. Tayler for checking the manuscript, but any mistakes that remain are my own.

Under the British Railways modernisation plan, a 'pilot scheme' was drawn up in 1955 by which a variety of diesel locomotives, engines and transmissions were ordered in small quantities. Among the 174 main-line units then ordered, in November 1955, were ten 2,300 bhp I Co-Co I diesel-electric locomotives. Powered by the Sulzer 12LDA28-A, twin-bank, four-stroke, pressure-charged diesel engine and using Crompton Parkinson electrical equipment, these ten locomotives were the first Peaks and the forerunners of a successful series of British diesel locomotives.

Numbered D1-D10 (now renumbered 44 001- 44 010), all the locomotives were built at Derby and delivered in 1959-60, using Sulzer 12LDA28-A engines built in Switzerland. Originally intended for both freight and passenger work on London Midland Region, the locomotives now only work freight trains and are shedded at Toton, near Nottingham.

The original specification gave the Class 44 a high starting tractive effort as it was intended that haulage of heavy freight trains of over 1,000 tons should be undertaken, as well as the working of express passenger trains. This specification was amply met, since the Class 44 units have a very high starting tractive effort of 70,000 lb., and a continuous tractive effort of 41,000 lb. at 16.5 mph. Each locomotive was named after a well-known English or Welsh mountain, giving rise to the "Peak" name applied to the whole series.

Subsequent to the original 'pilot scheme', orders for main line locomotives were placed and in 1957 a further batch of 127 Sulzer 12LDA28 engined I Co-Co I locomotives were ordered. Later designated Class 45, these use the 12LDA28-B engine, a development of the -A variant which by the addition of an intercooler, increased the power output to 2,500 bhp. Crompton Parkinson electrical equipment was again used, although slightly smaller traction motors were fitted, giving a maximum tractive effort of 55,000 lb., and a continuous tractive effort of 30,000 lb. at 25 mph. This is sufficient to start a 1,000 ton train on a 1 in 100 gradient.

Another later order was placed for 56 locomotives of the same I Co-Co I design and these Class 46 units were fitted with Brush electrical equipment instead of Crompton Parkinson as on the Class 44 and 45's. Brush equipment being preferred for maintenance reasons. The Sulzer 12LDA28-B was used, as in the Class 45, and, apart from the Brush electrical equipment, the Class 46 is in all other respects mechanically similar to Class 45.

MECHANICAL PORTION

Class 44, 45 and 46 locomotives are double nose ended, of full-width body design, and are carried on two eight-wheel bogies, each bogie having three driving and one carrying axle giving a I Co-Co I configuration. To improve the cab lines a nose extension was fitted, this extension forming a convenient housing for auxiliary equipment—in particular a blower at each end to supply air cooling to the traction motors.

The main structure of the underframe comprises two continuous side channels, between which is a central sub-frame to carry the power unit. A central box member runs from the sub-frame to each end of the locomotive to join the side frames which are curved to join at the centre

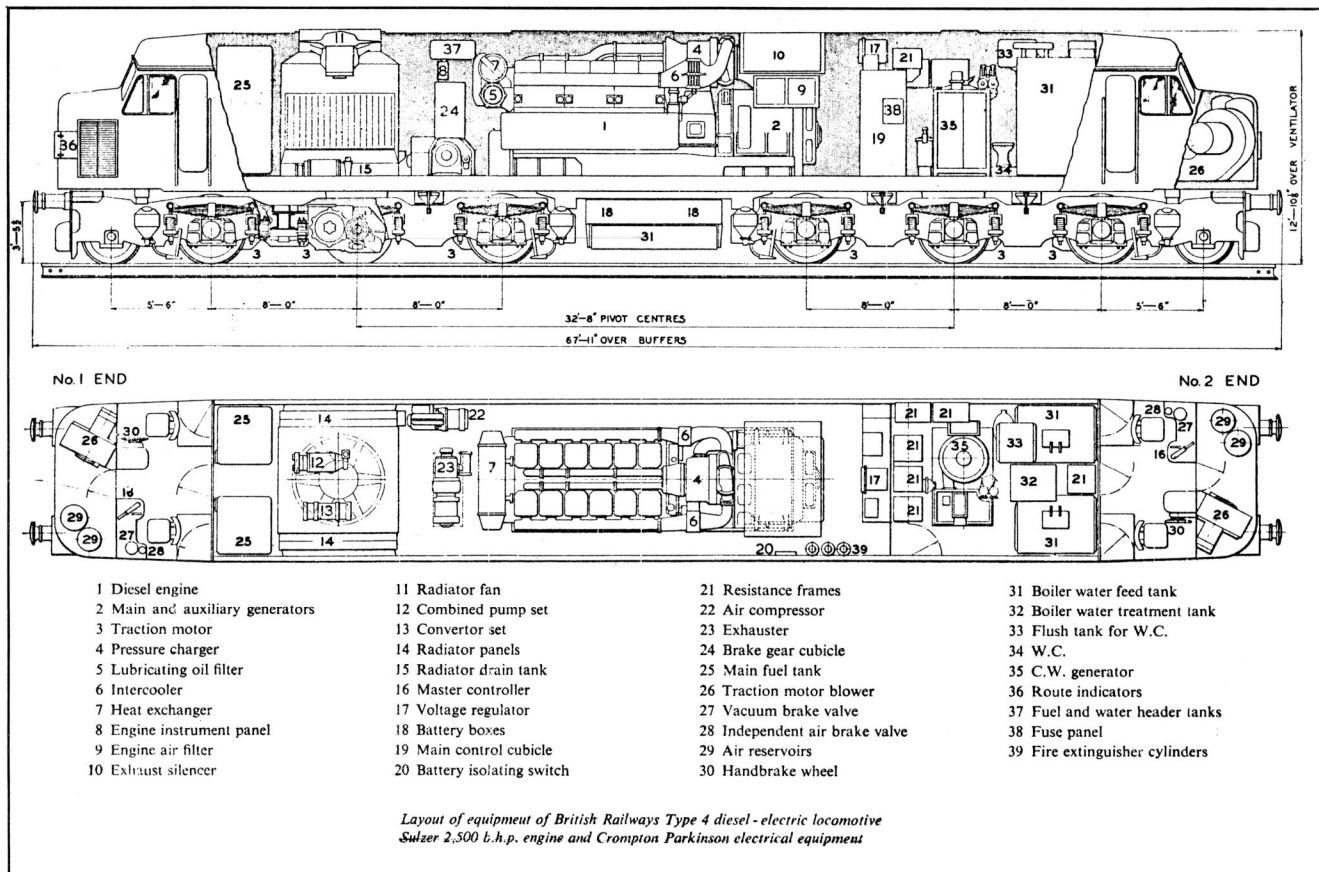

1 Diesel engine	11 Radiator fan	21 Resistance frames	31 Boiler water feed tank
2 Main and auxiliary generators	12 Combined pump set	22 Air compressor	32 Boiler water treatment tank
3 Traction motor	13 Convertor set	23 Exhauster	33 Flush tank for W.C.
4 Pressure charger	14 Radiator panels	24 Brake gear cubicle	34 W.C.
5 Lubricating oil filter	15 Radiator drain tank	25 Main fuel tank	35 C.W. generator
6 Intercooler	16 Master controller	26 Traction motor blower	36 Route indicators
7 Heat exchanger	17 Voltage regulator	27 Vacuum brake valve	37 Fuel and water header tanks
8 Engine instrument panel	18 Battery boxes	28 Independent air brake valve	38 Fuse panel
9 Engine air filter	19 Main control cubicle	29 Air reservoirs	39 Fire extinguisher cylinders
10 Exhaust silencer	20 Battery isolating switch	30 Handbrake wheel	

Layout of equipment of British Railways Type 4 diesel-electric locomotive Sulzer 2,500 b.h.p. engine and Crompton Parkinson electrical equipment

of the nose. Cross members carry the segmental bearings. Auxiliary cross members transmit weight to the front of the bogies through a spring pad at each side. The body side frames are built up with diagonal and vertical cross members between the underframe side channels and heavy cant-rail section. They run the length of the locomotive between the driving cab compartments. Two roof cross-members tie the side frames together. Sheet steel, lined with Navy-board for sound insulation is used for the body sides and roof.

Ventilation louvres extend the full length of the power compartment. There are no windows or side access doors, but a small access cover is provided at each side of the engine compartment. Over the power unit the roof is detachable as a complete section, and access covers are provided in the sides and in the roof for the servicing or removal of auxiliary equipment, boiler and electrical gear. An external recess in the roof accommodates the engine silencer, which is lagged with asbestos for heat insulation. The engine compartment floor is shaped round the engine sump and an oil sealing plate is fitted the full length of the locomotive between cab bulkheads to prevent oil reaching the traction motors and running gear on the underside of the frame. Any leakage is piped to a tank which is drained via a cock at the side of the locomotive. All piping running the length of the engine room is fitted between decks on one side of the locomotive, and all cables are laid in a sealed duct on the opposite side.

The three driving and one carrying axles of the three Classes have outside plate frames, based on the original Southern Region design for locomotives Nos. 10201-03. The load is transmitted to the bogie by the segmental bearings, and spring pads over the non-driving axle. This axle has bearings in a cannon box and forms a pony truck within the bogie. Two pairs of links with spherical bearings, mounted on the outer headstock, give movement of the pony truck about a vertical point near the adjacent axle. The truck is controlled by side springs and loaded by spherical seated springs. A Tecalemit axle-driven pump is fitted for pivot and linkage lubrication on Class 44, whilst on Classes 45 and 46 a Wakefield type DL pump is fitted.

The bogie frames are of welded and riveted construction. The main springs are laminated and the auxiliaries are coiled with rubber blocks. Springing is applied directly to the roller-bearing axle boxes, Timken in Class 44 whilst in Classes 45 and 46 both Timken and SKF types are fitted. The dragboxes are mounted directly on the bogie head-stocks with standard centre draw-gear and Turplat side buffers.

Spring-loaded side bearers are fitted on the dragboxes. Side movement is allowed on the centre driving axle and the traction motor moves with the axle. The normal limited side clearance is allowed on the other driving axles. Rolled-steel disc wheels and tyres (which are shrunk on) are fitted. The Smith's mileage and speed indicator is driven from the centre axle of one bogie.

Davies & Metcalfe Oerlikon-type brake equipment is fitted to all three Classes of the ''Peak'' family. Straight air braking is provided for the locomotive and vacuum brake operation for the train. With the vacuum brakes coupled, the locomotive air brakes are controlled by the driver's vacuum brake valve in the normal manner. Two brake blocks are operated through enclosed linkage by individual cylinders, for each driving wheel. The 6-in. diameter cylinders are mounted on the bogie frame cross members, and have integral slack adjusters. The equipment includes a locomotive brake release to facilitate uncoupling and an anti-slip brake to control wheel slip which reduces the need to use sanding to increase adhesion with heavy loads or with greasy rail conditions.

Air for the locomotive brakes and controls is supplied by an Oerlikon compressor, and the vacuum brake operation is by an S.L.M. type VL30 exhauster. Two $4\frac{1}{2}$ cubic ft. capac-ity air reservoirs are mounted in No. 2 end nose compartment and at No. 1 end there is a $5\frac{1}{2}$ cubic ft. auxiliary reservoir plus a 4 cubic ft. control air reservoir.

Cooling air is drawn, via the louvres in the bodyside, through the radiator panels of the Serck unit by a roof-mounted fan. This is driven by a two-speed 21 hp Crompton (Classes 44 and 45) electric motor with thermostatically controlled speed regulation giving speeds available of 1,320 and 785 rpm. On the Class 46, Brush variable-speed hydrostatic equipment is provided. Coolant tanks are fitted below the cooling panels to allow these to drain to prevent freezing when the engine is shut down or the radiator by-passed. Coolant is circulated by the motor driven pumping set through the lubricating oil heat-exchanger, engine, turbo-blower and radiator. The radiator is by-passed by a thermostatically-controlled valve when the jacket water is below the minimum working temperature. In the jacket the water is slightly pressurised to prevent cavitation. As on previous Sulzer engine installations, a motor-driven combined pumping set is used for coolant circulation, fuel oil transfer, and lubricating oil priming. The Sulzer pumps are driven by a Crompton or Brush motor running at 2,400 rpm.

ENGINES

The Sulzer 12LDA28-A and -B diesel engines are of the twin-bank, four stroke, exhaust pressure charged, inter-cooled (12LDA28-B only) direct injection types with a bore of 280mm and a stroke of 360mm. The 12LDA28-A fitted in Class 44 has a continuous rating of 2,300 hp at 750 rpm, and the 12LDA28-B fitted in the Class 45 and 46 locomotives is rated at 2,500 hp at 750 rpm. When the "Peaks" were introduced in 1959, the 12LDA28 was the most powerful railway diesel engine in Western Europe.

The cylinders are arranged in two vertical banks of six, each bank with its own crankshaft driving a common output shaft through straight spur gearing. This has a step-up ratio of 1:1.44, giving a generator speed of 1,080 rpm at 750 rpm engine speed which permits both a main and auxiliary generator to be used. Piston speed in the 12LDA28-A is 1,770 ft. per minute and bmep 150 lb. per sq. in. with a charging pressure of 10 lb. per sq. in. above atmospheric. Piston speed in the 12LDA28-B is 1,770 ft. per minute and bmep 165 lb. per sq. in. with a charging pressure of 16 lb. per sq. in. (1.1 atm) above atmospheric.

The line components of the 12LDA28-A and -B engines

(pistons, liners, connecting rods and bearings) are identical and interchangeable in the corresponding variants of the Sulzer 6LDA28 and 8LDA28 engines fitted in BR Classes 24, 25, 26, 27 and 33.

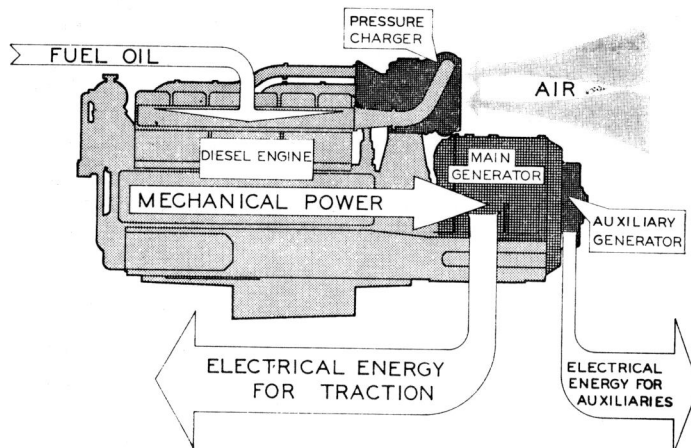

The process which takes place in the Sulzer 12LDA28A diesel engine and Crompton Parkinson GC426-A1 generator. (Sulzer Bros. Ltd.)

In the 12LDA28-A an exhaust-gas pressure charger is fitted but developments by Sulzer Brothers on this engine produced the 12LDA28-B which has an intercooler fitted after the pressure charger for increased power output. The cooling of the air after it has passed through the pressure charger is achieved by an air-to-water intercooler. This is a single-flow cooler with round tubes in which the secondary surface takes the form of flat fins, the latter enabling the whole volume within the cooler to be occupied by active surface. The tube stack is fitted directly into a wide portion of the induction manifold between the pressure charger and the engine, thus making the intercooler an integral part of the engine. Cooling water is derived from the main engine cooling circuit which eliminates the need of a separate water circuit with additional piping, radiators and pumps.

The cylinder block is of the wet liner type with a heavy section top plate and a central welded-in inlet water pipe. A single camshaft on the outside of each bank is used for the operation of the valve gear and fuel injection pumps. Three fibre glass inspection panels and one full-length

steel cover on each side of the engine give access to fuel pumps and crankcase respectively.

The crankcase is an integral structure made up of a series of transverse members welded to mild steel fabricated longitudinal plates. A cross member is fitted at each main bearing position and also at the output gear bearing positions. The box-form side girders are extended to provide a mounting for the generator. The bottom is closed by the sump to complete the crank members. Large diameter oil pipes are welded in and the transverse members are carried up to cylinder liner level to form the attachment facings for corresponding members in the cylinder block fabrication.

The two crankshafts are heat-treated alloy steel, fully machined with balance weights and have elliptical crank webs. The journals are finish ground and hollow bored to reduce weight.

Main bearings are made of tri-metal, steel backed, copper lead lined, with a soft permanent running surface to give an extremely reliable component. A Holset vibration damper is fitted to the free end of the crankshaft so that there is no limitation of the choice of engine speed over the full working range.

Connecting rods are of the H-Section nickel chrome forged steel type and are drilled in the longitudinal axis to carry lubricating oil from the big ends to the small ends and, for cooling purposes, to the piston. The big end bearings are similar to the main bearings whilst the small end bearings have a steel back with a layer of copper lead.

The pistons are made of aluminium alloy and are cooled by the lubricating oil from the engine system fed from the fully floating gudgeon pin. Individual cylinder heads are made of alloy cast iron with an open combustion space and porting to permit the use of only one inlet and one exhaust valve per cylinder, with the injector fitted centrally. The camshaft is gear driven from the crankshaft and no chain drives are used in the engine. Valves are operated through push rods and rocker arms and all moving parts including the valve tappets are automatically lubricated, with identical inlet and exhaust valves.

The ten engines of the Class 44 were built by Sulzer Bros. at their headquarters at Winterthur. The 183 engines installed in the Class 45 and 46 locomotives, together with 7 spare units, were built by Vickers-Armstrong Ltd. at Barrow under sub-contract from Sulzer Bros. This was for

On the now-closed Peak Forest route from Manchester to Derby, Class 44 D1 *Scafell Pike* at *Millers Dale* with an up express for St. Pancras. Use of the first ten 'Peaks' on passenger trains was discontinued when the more powerful variant (Class 45) became available in quantity.

[*Sulzer Bros. Ltd.*]

economic and productivity reasons and at one time Vickers were building a complete engine per working day.

There have been two modifications on the 12LDA28 engines fitted in a Class 44 locomotive (D2 *Helvellyn*) and in a Class 45 (D57). The 12LDA28-A in D2 was uprated from 2,300 hp to 2,500 hp by the fitting of a simple charge air intercooler of the type used in the 12LDA28-B. The engine was set to run at a 10% overload rating and so this experiment only ran for a short while as this uprating would mean a faster rate of wear on the components. The use of the intercooler on D2's engine was the forerunner of the 12LDA28-B fitted in the Class 45 and 46 locomotives.

7

Class 45 D57 was modified in May 1962 at the Barrow works of Vickers Armstrong, the original 12LDA28-B engine being modified to 12LDA28-C specification with the fitting of the more advanced intercooler, pistons, connecting rods and governor used in the Class 47 C engine. This modified engine was rated at 2,750 hp and the locomotive ran in this form from May 1962 until the early part of 1964 when it was converted back to 12LDA28-B specifications.

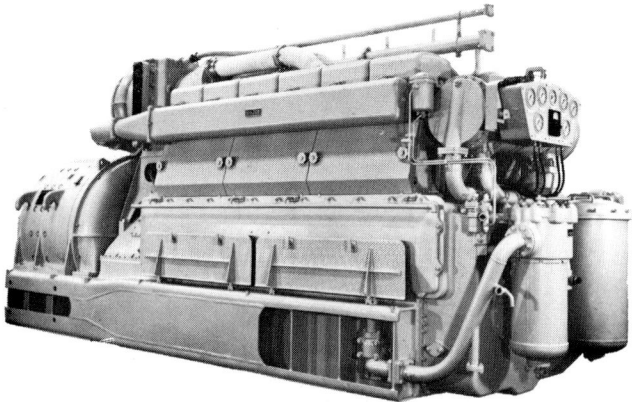

ELECTRICAL EQUIPMENT

The combined main and auxiliary generator assembly is carried on an extension of the engine crankcase. The armatures are mounted on to a bottle-shaped cast-steel rotor, bolted to the engine coupling flange and carried at the rear end in a single roller bearing. Classes 44 and 45 are fitted with Crompton Parkinson GC426-AI main generators which are self-ventilated 10-pole machines with a continuous rating of 1,531 kw, 580 volts, 2,640 amp. at 1,080 rpm. The auxiliary generators are eight-pole, self-ventilated machines with a continuous rating of 90 kw 220 volts 410 amp. and the voltage is held constant at all speeds between 650 and 1,080 rpm. To reduce overall length, the auxiliary generator is recessed into the field system of the main generator, the commutator of which is at the driving end. Compensating windings are fitted to the main generator poles faces and any field coil-self field, separate field, de-compounding and starting field – can be removed and replaced with the compensating winding in position. Class B insulation is used throughout.

The electrical equipment fitted in the 56 locomotives of Class 46 was supplied by Brush Electrical Engineering Co.,

this being (as already noted) the only difference in electrical equipment from those of Class 45. The Class 46/12LDA28-B engine is fitted with a Brush TG 160-60 main generator, which is a self-ventilated, 10-pole machine with a continuous rating of 1,533 kw, 730 volts, 2,100 amp. at 1080 rpm. The auxiliary generator is an eight-pole, self-ventillating machine with a continuous rating of 50 kw, 110 volts.

Maintenance of brushes in the Classes 44 and 45 is facilitated by the fitting of Crompton Parkinson rotating brush gear. There is a high proportion of common parts fitted on the main and auxiliary generators of these and other Crompton Parkinson-equipped locomotives. A toothed ring is incorporated in the generator fan to enable the engine to be rotated during servicing. Cooling air for the generators is drawn from the engine room and discharged through the floor.

The six traction motors of the Class 44, Crompton Parkinson-type C171-B1, are series-wound and force-ventilated with an individual continuous rating of 305 hp, 440 amp, 580 volts, and a one-hour rating of 305 hp, 485 amp, 530 volts. These motors are similar, apart from the gear ratio, to those fitted in the Class 26 and 33 Sulzer/Crompton Parkinson/Birmingham R.C. & W. Locomotives of 1,160 and 1,550 bhp. The motors are axle hung and have chevron-shaped rubber nose-suspension units on the two outer motors to limit side movement. On the centre motor of each bogie sandwich-type rubbers, which allow the motor to float with the axle, are fitted. Wiseman resilient-type gear wheels are fitted in the single reduction gearing, and rotatable brush gear is incorporated to facilitate pit servicing.

Air for the force ventilation of the motors is supplied by two Aerex Hyperform blower units. Each blower is driven by a Crompton two speed 11.5 hp motor at 1,180 or 2,360 rpm and has a maximum delivery at 6,750 cubic ft. per minute. Air for the blowers is drawn through louvres in the machinery compartment side with one blower floor mounted in each nose compartment, delivering air to the three motors of the adjacent bogie via the box-form longitudinal frame member with air branches for each motor. Current for the traction-motor blowers, compressor, exhauster and pumping-set motors is supplied at 220 volts. For the control circuits and lighting supply a rotary converter is used to reduce voltage from 220 to 110 volts.

The six traction motors fitted to Class 45, Crompton Parkinson type C172-AI, are slightly smaller than those of

Crewe-built D69 when new outside the works in 1960, showing the original form of fixed nose ventilator for the traction motor blower.

[*Sulzer Bros. Ltd.*]

Class 44, but are of similar construction and are axle-hung in exactly the same manner as for the first ten locomotives, although David Brown gear wheels are fitted. The C172-AI traction motor has an individual continuous rating of 337 hp, 445 amp, 615 volts, and a one-hour rating of 340 hp, 485 amp, 580 volts. Ventilation of the traction motors is supplied by two Sturtevant blower units.

The six traction motors of the Class 46, Brush type TM 73-68 Mk.III are nose suspended having continuous rating of 327 hp apiece at 350 amps, and 730 volts.

CONTROL EQUIPMENT

Allen West control gear is housed in a dust-tight cubicle near the generator. Cam-operated contactors are used for engine starting, traction motor switching and field control. The motor reverser is electro-pneumatic and the auxiliary equipment contactors are electro-magnetic. Links are provided for the emergency isolation of individual motors.

The master controller, through the engine governor, provides full control of the engine speed and electric power output. This, coupled with carefully chosen characteristics, enables tractive effort control to be achieved without the use of notching contactors. The engine governor is loaded by air pressure controlled by an air valve operated by the controller power handle in the driving cab. The load regulator is driven by the engine governor and comprises an oil

D100 Sherwood Forester *at Derby in mint condition after the naming ceremony. This Class 45 was built by Crewe and entered service in 1961, but was the first of the Class 45s to receive nameplates.*   [*Sulzer Bros. Ltd.*]

servo-motor, incorporated in the engine governor, driving a flange-mounted rheostat adjusting the generator excitation, so that the electrical load balances the pre-determined engine output at any selected speed. A pressure-charging protection unit limits the fuel injected according to the amount of air available from the pressure-charger.

The traction motors are connected in permanent parallel across the main generator and five stages of motor field weakening are automatically controlled by the load regulator and a current relay. The load regulator controls the main generator field so that engine power remains constant at the setting selected by the power handle irrespective of locomotive speed, as well as operating contacts in the control gear for diverting the traction motor field current to obtain high speed running.

A lead-acid battery of 96 cells and 142 amp.-hr. capacity is carried in four pull-out boxes suspended from the underframe. External charging sockets, lighting connections and inspection lamp sockets are provided on each side.

LOCOMOTIVE LAYOUT

The large power compartment contains (starting from No. 1 end) two 400 gallon fuel-oil tanks, radiator unit, oil and water pumping set, converter set, air compressor, vacuum exhauster, engine and generator group, and electrical control cubicle. The compartment has large bodyside louvred apertures through which air for engine, electrical and auxiliary equipment is drawn. Engine air is taken via Vokes metal type air filters mounted in a filter box on top of the main generators.

The Serck radiator is adjacent to the two vertical fuel tanks which are located against the cab rear bulkhead at No. 1 end. Apertures are cut in the radiator fan cowling for the extraction of hot air from the compartment. Access from the cab to the compartment is by a walkway between the tanks and through the radiator tunnel.

Mounted on a steel frame over the exhauster is a cubicle containing the brake control valves. Above this cubicle is a 40 gallon auxiliary fuel tank and a 15 gallon engine cooling water header tank. The electrical control cubicle extends across the compartment at the generator end at one side of a communicating doorway through the bulkhead into the boiler compartment. Switches and instruments which the driver may be required to observe or operate are contained in a separate floor-mounted cubicle.

The train heating boiler fitted originally in Classes 44, 45 and early 46s, is a fully automatically oil-fired stone-Vapor Unit (Type OK4625) with a generating capacity of 2,750 lb. of steam per hour at 70 lb. per sq. inch. Water for each boiler (where retained) is carried in two 520 gallon tanks, mounted vertically against No. 2 end cab bulkhead and a further 300 gallon tank is underslung between the bogies. All tanks and water pipes are galvanised and the T.I.A. system of feed-water treatment, incoparating a hydrostatic doser tank, is fitted. Both the air intake and the boiler exhaust are ducted to the roof.

The boilers and water tanks in the Class 44 were removed in 1962-63, the Class 44's then being confined to freight traffic only, there being sufficient locomotives of Class 45 for passenger workings.

In the Class 45 the boiler and water tanks remain in some locomotives (reclassified 45/0 Nos. 45 001-077) and the remainder now are, or will be, classified 45/1 (Nos. 45 101-150) and are fitted with electric train-heating equipment only, their boilers and water tanks having been removed.

The Class 46 boilers and water tanks remain as standard fitting, although ultimately they will be fitted with electric train-heating equipment.

Each of the two driving compartments contains identical equipment, with the driving position on the left side and the second man's seat at the right. On the right side is located the hand wheel for the brake, and a panel with a trainpipe steam pressure gauge and boiler warning light in the Class 45 and 46 locomotives fitted with steam-heating boilers. An electric cooker is positioned beside the centre gangway.

The driving position (see diagram) has the driving controls located on a desk, with instruments and warning lights, grouped on a vertical panel. The 'deadman's' switch is treadle operated. At the right of the driver's seat is the master controller, which has a button in the power handle for operating the anti-slip brake. On the left are the straight air and vacuum brake controls. The instrument panel is fitted with a vacuum gauge, automatic air brake gauge reservoir, and straight air brake gauge, ammeter and speedometer. The brake release button and the usual three warning lights for 'Engine stopped', 'Wheelslip' and 'Fault' are above the instruments. Lighting switches are grouped on a roof mounted panel positioned above the driver. On the rear bulkhead are the fire extinguisher controls and dial gauges showing contents of fuel and water tanks.

The first of the Class 45s, D11, was built at Derby in 1960. After a good start the performance of the Class 45s was for some years disappointing but from 1967 onwards became notable for power and consistency. In blue livery with D prefix painted over, 11 pauses at Carlisle in March 1973, having worked a steam-heated train over the Settle & Carlisle line. [H. L. Ford]

Class 46 D139 (built at Derby), in February 1962, showing the original form of headcode display carried by D138 – D166, but latterly borne only by some Class 45 locomotives. The absence of the angular ventilation panel for the enlarged main generator on this, the second Class 46 to be built, is an interesting point of detail compared with later photographs such as on page 22 and page 27. [Sulzer Bros. Ltd.]

Certain warning and safety devices are fitted in the 'Peaks' and faults are detected by four indicator lights, red, amber, blue and white. These show a dim light when correct and when a fault occurs the light becomes bright. The red light indicates that the engine has either stopped or failed to start. The engine is shut down automatically in the event of low lubricating oil pressure, lack of cooling water pressure and if the engine overspeed device is tripped. The amber light indicates that one or more pairs of wheels are slipping, the failure of a traction motor or the locomotive's handbrake is on. Power is cut off from the traction motors and the engine speed reduced to idling if an earth fault occurs or the locomotive loses air pressure or brake vacuum.

The blue light indicates high water temperature, loss of air pressure or brake vacuum, earth fault relay tripped or a traction motor blower failure. Confirmation of the latter is indicated by a red light on the engine room control cubicle. Finally, in those locomotives still fitted with a steam-heating boiler the white light goes out to indicate that the boiler is switched off or an indicator circuit has failed.

Fire detection switches, which operate the cab warning bell, are positioned at suitable points. Operation of the built-in fire extinguishers is by the cab control or externally at ground level. Three 50 lb. CO_2 cylinders are located in the power compartment and three portable extinguishers in each cab.

The driver's and second man's upholstered seats are adjustable and a sliding window in the cab side is adjacent to each seat. Electric heaters are positioned in a $\frac{3}{4}$in. space between the glass of the double-glazed driver's windscreens for de-misting and de-frosting, and Trico-Folberth pneumatic wipers are fitted. Cressall cab heaters of 500 watts each are mounted on the bulkhead beneath the seats and beside the crew positions to give a total of 4 KW per cab controlled by three separate switches. Two combined inlet and extractor ventilators are fitted in the roof of each cab.

Cab layout and controls of the Class 45/46 [Sulzer Bros. Ltd.]

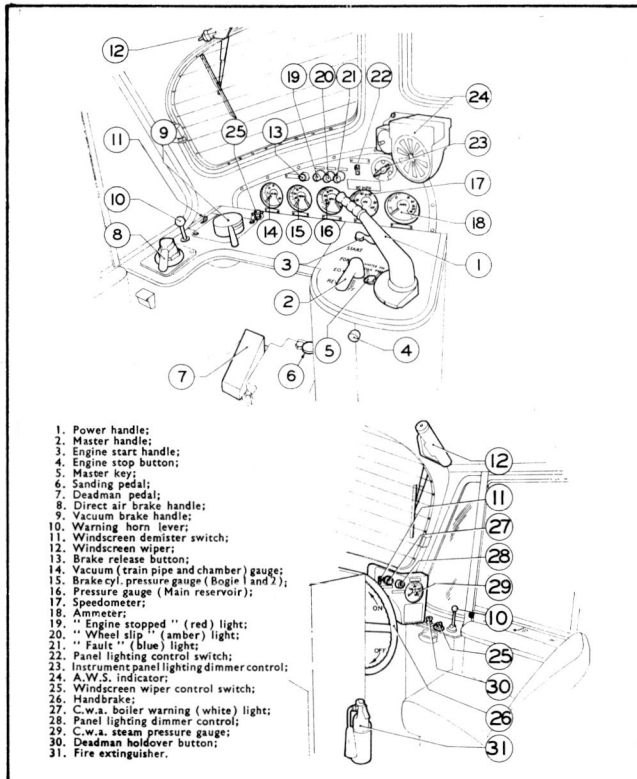

1. Power handle;
2. Master handle;
3. Engine start handle;
4. Engine stop button;
5. Master key;
6. Sanding pedal;
7. Deadman pedal;
8. Direct air brake handle;
9. Vacuum brake handle;
10. Warning horn lever;
11. Windscreen demister switch;
12. Windscreen wiper;
13. Brake release button;
14. Vacuum (train pipe and chamber) gauge;
15. Brake cyl. pressure gauge (Bogie 1 and 2);
16. Pressure gauge (Main reservoir);
17. Speedometer;
18. Ammeter;
19. "Engine stopped" (red) light;
20. "Wheel slip" (amber) light;
21. "Fault" (blue) light;
22. Panel lighting control switch;
23. Instrument panel lighting dimmer control;
24. A.W.S. indicator;
25. Windscreen wiper control switch;
26. Handbrake;
27. C.w.a. boiler warning (white) light;
28. Panel lighting control;
29. C.w.a. steam pressure gauge;
30. Deadman holdover button;
31. Fire extinguisher.

ALLOCATION

The first Class 44 locomotive, DI *Scafell Pike* entered service in 1959 and was allocated to Camden shed, spending a short time at Longsight (Manchester). Later on, D2 *Helvellyn*, D3 *Skiddaw* and D4 *Great Gable* were allocated on introduction into service to their birthplace at Derby. After their original allocations to Camden and Derby, all ten Class 44 locomotives moved to Toton, and have stayed there ever since.

All the early Class 45 'Peaks' were allocated to Leeds (Holbeck) and many still remain there. 76 Class 45's are allocated to the Nottingham Division and ten to the London Division.

A good view of the roof detail showing the location of the main radiator extractor fan at the No. 1 end (nearest camera), and the two insets at the No. 2 end for replenishment of train heating boiler water tanks. D166 seen here at Newcastle in March 1971, was built at Derby in 1962.

[D. E. Canning]

A close-up of the No. 2 cab end of D9 Snowdon, one of the ten true 'Peaks', in rather work-worn condition at Toton depot in April 1970. Snowdon and Tryfan are the only two 'Peaks' with the vertical-slatted grilles illustrated. In addition, Snowdon has received a Class 46-type single-piece headcode panel at the No. 1 cab end as a retrospective replacement item.

Class 46 locomotives are divided amongst Western and Eastern Regions. Of the 32 on Western Region, 8 are at Laira and 24 at Bristol Bath Road. Plymouth Laira later on had the first four (D138-D141) transferred there. All 24 Eastern Region Class 46's are at Gateshead.

Although Class 44 'Peaks' only work freight trains in the Nottingham Division, Class 45's and 46s work passenger and freight trains over many areas of the Midland, Western and Eastern Regions. Apart from their dominance on the St. Pancras-East Midland (Leicester, Derby, Nottingham, Sheffield and Leeds) services, as well as 'The Thames-Clyde Express' over the Settle and Carlisle line, the 'Peaks'

work many of the North of England (Liverpool, Manchester, Leeds and Newcastle) to the West Country (Bristol, Paignton, Plymouth and Penzance) services. 'The Cornishman' and 'The Devonian', from Leeds to Penzance and Paignton respectively are a 'Peak' stronghold.

The 'Peaks' are big and heavy, and in many ways similar to the English Electric Class 40 locomotives, although the latter are less powerful. When conceived under the 1955 Modernisation Plan, twenty years ago, and on their introduction in the late 1950's they were the most powerful single engined diesel locomotives in the world. They owe much of their looks (as do the EE Class 40's) to the two LMS

C-C diesel-electric 1,600 hp locomotives Nos. 10000 and 10001 introduced in the late 1940's, the first main line diesel locomotives in British Rail service.

Many of the 'Peaks' are named and the class name commonly given to the whole series of 193 locomotives arises from the decision to name the first ten after mountains of England and Wales. The Class 44s stand out amongst all the diesel-electric classes on BR because of their evocative and well chosen names. The first four have been mentioned earlier, and the others came out of Derby into the public eye with similar resounding names; D5 *Cross Fell*, D6 *Whernside*, D7 *Ingleborough*, D8 *Penyghent*, D9 *Snowdon* and D10 *Tryfan*. Twenty-five Class 45s and one Class 46 were named after British Regiments, whilst D60 is an odd one out with the name *Lytham*

St. Annes. This follows a Midland tradition, using names formerly carried by LMS 'Royal Scots' and 'Patriots' as indicated in Appendix 2.

LIVERY

All Class 44, 45 and 46 locomotives were introduced in standard B.R. Brunswick Green Livery. A grey stripe ran from bottom of the cab door from each end of the locomotive and in addition the long horizontal ventilation louvre on either side of the locomotive was painted grey similar to this band. The B.R. 'Lion' crest transfer was located on the No. 2 end left-hand side a few feet from the cab door on the same horizontal axis as the door handle and similarly on the right-hand side of the locomotive. Nameplates were located between the small ventilation

Class 44 D7 Ingleborough *working one of the heavy coal trains for which the higher tractive effort makes the Class 44 most suitable. Retaining Swiss-built 2,300hp engines, all ten of the first 'Peaks' have spent the major part of their working lives operating from Toton on duties of this kind.*

louvre and the large horizontal louvre near the No. 2 end on either side. The front of the bogie was painted bright red. In late 1962 a yellow warning panel was painted on the nose of all the locomotives beneath the headcode indicator (originally painted on the later Class 46 locomotives). The roof was painted grey.

In late 1967 B.R. blue was gradually applied to all Class 44, 45 and 46 locomotives, with an all yellow nose.

The locomotive number was originally located underneath the cab side window on all four ends of the locomotive. This had the standard D prefix until 1968, when steam locomotives were finally withdrawn from service, although there were one or two of the 'Peaks', still with the letter D prior to renumbering. Beneath the number on two opposite sides of the locomotive is a rectangular works plate stating either B.R. Derby or Crewe built, and year of construction.

HEADCODES

Class 44 locomotives were fitted with B.R. disc headcodes (three along the base of the nose and one in the centre at the top), for the standard train identification method before four character headcodes were introduced and was used to identify any of the ten standard classes of trains. When two sections of the disc are opened out flat a white circle with a red coloured spot on the lower half is exposed. All the Class 44 locomotives will probably be converted to four character headcode panels as several have already—a pity as the discs are another distinguishing feature of the first 'Peaks'.

The gangway doors in the nose of the Class 44 locomotives were fitted for multiple operation, although this was not used in service.

Class 45s were originally delivered with a four character headcode divided into two in square panels on either side of the nose. A gangway door similar to the first 10 'Peaks' was fitted to several of the early class 45's but this was short lived. This was in preparation for multiple operation and jumper cables were fitted on all locomotives, but were all subsequently removed after this type of operation was regarded as being unnecessary. On all the Class 46's, a four character headcode panel in the centre of the nose was fitted.

Later, when new nose sections were fitted to the Class 45 and 46 locomotives, because of damage, etc., a variety of headcodes have appeared. Apart from the original two-panel, four-character headcode, the Class 45's have also been fitted with a single four-character headcode panel in the centre of the nose. This was either a complete panel or was divided into two separate but closely placed panels, both these types being fitted on the Class 46.

There are several 'odd men out' in the 'Peaks' headcodes, whereby a different type of headcode and new nose section has been fitted at one end of the locomotive after frontal damage, details of which can be found in Appendix 3.

EXTERNAL FEATURES

All Class 44, 45 and 46 locomotives are externally of the same design, but there are several distringuishing features which differentiate each class and the small sub-divisions in the same class.

Class 44s are primarily identified externally by their headcode discs—although these are now being removed in favour of four character panels—and the access doors in the nose. As mentioned in the headcode section there are several types of headcode panels fitted on Class 45 locomotives, two separate square panels on either side of the nose, or a single four character panel in the centre. (See photographs of the various locomotives which illustrate these differences.) Class 46s are all fitted with the single four character panel, divided into two different types as already described, similar to some of the Class 45's.

There are also differences in the 'Peaks' in the ventilation louvres along the bodyside. All three Classes have a rectangular louvre for the traction motor blower, on the side of each nose. Originally the ventilation louvres on the 44, 45 and 46s were bolted to the side of the nose, but in Classes 45 and 46 they were later fitted with hinges to allow access to the air-filters which were installed later. Next from the No. 1 end of the locomotive (moving right) is a large, nearly square-shaped, ventilation louvre which is followed by a long rectangular louvre nearly 26 ft. long and 3 ft. wide. on the upper half of the bodyside. These two louvres are identical on all three classes of the 'Peaks', although those of 44009 (D9) *Snowdon* and 44010 (D10) *Tryfan* have the same shape but the grilling runs vertically from supporting lengths of metal interwoven between them.

Beneath the long ventilation louvre provision was made

Only the Class 44s have retained nose access doors for crew changing, although no longer used for that purpose. Renumbered No. 44 004 Great Gable, stands alongside D39 of Class 45, in Lloyds sidings, Corby Steelworks. [F. R. Kerr]

LAIRA   140

AB-VB	Class	46
	Weight tons	139
	Brake force tons	63
	RA	7
	Max speed mph	90

BRITISH RAILWAYS
DERBY BUILT

*Two views at Penzance of Derby-built Class 46 No. 140 after transfer from the Midland to Western Region. Class 46s require a slightly different driving technique from the Class 44 or 45, involving more progressive power application, but has **proved** popular with crews for its reserves of strength to cope with heavy loads.*

[H. L. Ford]

for a small angular louvre, marked by a panel which, if needed, could be removed for an additional ventilation area. All the Class 44s and many of the Class 45s and 46s still have this panel fitted, but a few 45s and numerous 46s have a louvre fitted in this space.

Another ventilation louvre was fitted on all the "Peak" locomotives some 5 ft. along from the large horizontal louvre towards the No. 2 end. All named locomotives have their nameplates located between these two louvres. This final louvre is small and vertically placed on the bodyside. It was fitted identically (except for D9 and D10 mentioned above) on all three classes, but was later removed and a metal panel placed over the space in many of the Class 45 locomotives. The British Railways "Lion" Crest transfer was located between the small louvre and the cab door.

Looking at the locomotives in plan view, three footholes and rungs leading to the roof, located just behind the door on the No. 1 end left hand side and on the No. 2 end right hand side, were fitted on all the "Peaks", but were subsequently covered up by metal plates.

PERFORMANCE

The most important test for any locomotive is the way in which it performs in service. As mentioned previously, the "Peaks" have seen service over many lines in Britain working both passenger and freight trains. During the last fifteen years they have established themselves as some of the best Type 4 diesel-electrics in service, and to many observers the Class 45 is the best of the three variants. Over all three Regions where they are allocated on passenger duties, Classes 45 and 46 show consistency and power, as exampled on the major routes detailed below.

(a) North East—South West Services. At the present time this route affords the best comparison of the Type 4's which work services approximately in the following proportions:

Leeds (Holbeck) Class 45	40%
Nottingham (Toton) Class 45	15%
Bristol and Plymouth Laira Class 46	20%
Cardiff Class 47	20%
Bristol Bath Road Class 50	4%
Gateshead Class 46	1%

Table 1. Approximate distribution of booked workings on NE-SW route.

Although workings are determined by loadings, with a tendency for Class 46s to be allocated the heavier trains. Class 47's mainly work diagrams out of Cardiff, but they make many appearances on other workings, in particular on summer Saturdays.

The most important place where power is needed on the line is ascending Lickey bank, towards Birmingham, up a gradient averaging 1 in 37, and with an approach speed restriction of 80 mph. From the runs shown in table 2 the

Class 45 shows itself the leader, its Crompton Parkinson electrical equipment being regarded as performing better than that of the Brush-fitted Class 46 or 47.

	Gross Load (Tons)	Bromsgrove Maximum (mph)	Blackwell Minimum (mph)
Class 45	360	78.5	33.0
Class 46	424	75.0	25.3
Class 47	370	74.3	28.0

Table 2. Average speeds of 18 unchecked climbs of the Lickey Incline.

(b) West Country Services. The Class 46's transferred to Bristol Bath Road and Plymouth Laira in recent years have mainly taken over "Warship" and "Hymek" workings. Unfortunately they are not able to match the Class 52 "Westerns", the latter being lighter, more powerful and having a higher tractive effort. The 46's tend to be more popular with crews than Classes 47 and 50s because of their good tractive effort from rest, demonstrating potential suitability to replace 'Westerns' on heavy freight trains over the switchback grades in South Devon and Cornwall.

	Class 52 No. 1071 Western Renown 10/344/370		Class 46 No. 46 004 13/446/460	
Par	00.00	—	00.00	—
St. Blazey Bridge	06.16	slack 7/27	04.30	32/35
MP 284	09.00	26	06.53	19.5
MP 285	11.25	24	10.18	17/16/5
Luxulyan	13.43	30	13.46	25
	(gradient 1 in 36/40)			

Table 3. Comparison of Class 52 and 46 Timings on the Newquay line

The Class 45's which work west of Bristol are usually from Leeds (Holbeck) on workings to Plymouth and back. On the fast sections between Taunton and Exeter, these trains are often worked very hard and Class 45s respond well to such treatment.

(c) Leeds Holbeck Class 45's. The Holbeck engines generally give the best performances of all the Class 45's, and thus of all the 'Peaks'. They also work extensively over the Settle-Carlisle line, which demands a great deal from the locomotives. An exceptionally good performance was put in by Class 45 No. 34, as shown in Table 6, a truly memorable experience.

Settle	00.00	—
Stainforth	03.07	50
Helwith Bridge	05.59	59/64
Horton	07.33	62/64
Selside	09.47	62/64
Ribblehead	12.07	63/64
Blea Moor	13.18	63
Tunnel South	14.06	63

Table 4. Class 45 log up part of the 'Long Drag' on the Settle and Carlisle line.

The number and severity of gradients on this line present a hard task for any motive power units. The $13\frac{1}{2}$ miles from Settle to the south portal of Blea Moor Tunnel are largely at 1 in 100, with shorter stretches at 1 in 176, 1 in 200 and 1 in 300, and only one small section running level. The maximum estimated drawbar horsepower for the Class 45 is 1900 hp.

An interesting feature of the 'Peaks' and shown up best of all by the Class 45's, is the ability—like the 'Deltics'—to run at very high speed for sustained periods of time. In this respect they can run them a very close second, a remarkable feat since the Class 55 weighs some 37 tons less and has some 800 bhp more available. Table 5 shows a Toton Class 45 and a Gateshead Class 55 on the straight and level East Coast Route between York and Darlington.

	Class 55 No. 9002 'The King's Own' Yorkshire Light Infantry 8/278/290		Class 45 No. 84 Royal Corps of Transport 8/270/290	
York	00.00	—	00.00	—
Skelton Junct.	03.17	59	03.21	58
Beningbrough	06.18	90/101	06.36	82
Alne	09.48	100/102	10.25	94
Pilmoor	12.44	98/103	13.32	96/94
Thirsk	16.25	97/96	17.24	97/98
Otterington	19.03	101	20.04	96/94
Northallerton	21.10	97	22.17	95/98
Danby Wiske	23.35	90/89	24.38	96/98
Eryholme	27.00	44/97	27.52	95/97/SIGS.
Darlington	32.37	—	34.06	(=33.00)

Table 5. Comparison of classes 55 and 45 on the East Coast Main Line.

(d) St. Pancras-Midland Main Line. This is probably the best route for the "Peaks" to show their form in numbers as they dominate the workings. The majority of the trains receive a Toton 45, although there are regular Holbeck workings, with 46's and 47's to supplement them. The light loadings and a maximum gradient of 1 in 119 combine to ensure a good performance from nearly every locomotive on this lightly-trafficked line. The superiority of the 45's as the best Type 4 is again shown in Table 6, the 47 not showing so well. Both trains are the 14.10 Sheffield to St. Pancras.

	Class 45 No. 25 8/268/285	Class 47 No. 1934 8/268/285
Kibworth North	85/83—2 mph.	84/81—3 mph.
Desborough North	49/66+17	58/67+9
Sharnbrook	73/70—3	78/70—8
Leagrave	90/85—5	97/83—14

Table 6. Comparison of class 45 and 47 locomotives up four successive climbs on the St. Pancras main line.

Amongst the best of the Toton locomotives are Nos. 82, 84 Royal Corps of Transport (See Table 7), 121 (mentioned previously) and 127 of which M. J. Oakley says "on it's day the best Type 4 in Britain". This locomotive has been recorded by Paul C. Edwards as once reaching 90 mph in the dip at Flitwick with 8 coaches on, from a stand at Bedford, a feat without parallel. A run on 6 July, 1974, recorded by M. J. Oakley, with No. 127 produced the following figures:

Class 45 No. 127 10/348/370

Derby	00.00	—
Repton	06.30	84
Clay Mills Junction	08.13	90/91
Burton	10.18	—
Whichnor Junction	05.52	79
Elford	08.09	85
Tamworth	11.00	90
Kingsbury	14.37	93/94
Water Orton	17.27	Slack 74/80
Saltby	21.48	66/SIGS. Stop
Birmingham	29.13	(=26.00)

Table 7. Class 46 log between Derby and Birmingham New St.

The St. Pancras-Midland Main line stands out amongst all the others—apart from the ER stretch between York and Darlington—as a route which enables locomotives and crews to give some very good times and high speeds.

Apart from the aforementioned, "Peaks" often work to many other parts of the country, mainly on secondary services between Newcastle and Edinburgh or relief trains from Newcastle to Kings Cross. More 45's are now to be seen in East Anglia, a regular diagram being to Cambridge. Bristol Bath Road 'Peaks' are occasionally seen at Shrewsbury on Euston trains whilst the Penzance and the Cardiff-Crewe parcels is usually 46 hauled. Freight workings include some into Southern Region territory.

The 'Peaks' work a wide variety of trains over many routes in Britain, and their performances on all of these are usually very good, and often excellent. Suffice it to say that these locomotives can be relied upon to do well on almost any job which is given to them. They appear to thrive on being worked hard and to improve progressively the harder they are used—a remarkable family of locomotives in many respects and a thoroughgoing credit to their builders, designers, maintenance staffs and crews.

They remain popular with crews for their ample power, and their ability to make up time when occasion demands it, although retaining among West Coast route enginemen on the LMR a reputation for being slightly rough-riding at certain speeds. After fifteen years' service they remain reasonably reliable and dependable machines, although less so than class 47s, with the prospect of many more years of service still ahead.

Class 45 No. 15 at full power with the up 'Cornishman' on 21 ▶ July 1970, near Dainton. The Class 45 responds particularly well to hard running such as this.   [G. F. Gillham]

The BR-Sulzer Type 4 and English Electric Type 4 (above) were designed to meet a similar specification though the EE Type 4 is the less powerful. Improved construction techniques and lighter bogies were adopted for the developed BR-Sulzer Type 4 produced later (below), which retained several of the internal equipment features of the 'Peaks'.

[J. R. Carter:
Peter J. Rose]

Loose-coupled heavy freights present few problems for the 'Peaks'; No. 46 032 passing through Selby station, July 1974
[Brian Morrison]

'THE PEAKS'—HEADCODE VARIANTS

Four types of headcode arrangement may be observed on locomotives of Classes 44, 45 and 46. The first ten locomotives (Class 44) were fitted with headcode discs and end access doors. The next 21 locomotives (Class 45 D11-D31) were introduced with two boxes for headcode blinds, the twin display panels being placed wide apart to allow space for access doors to be incorporated between them centrally. Class 45's D68-D107 were similar. All the remaining Class 45's were built with a pair of display blind panels set closely together. (Class 45 D32-D67; D108-D137). On these locos, it would have been possible to cut in access doors but this was not done, the idea of changing crews while the loco was moving having by then been abandoned. Nevertheless when Class 46 production commenced, the first 28 were also equipped originally with the pair of panels set close together centrally on each nose. (D138-D166). The final 27 Class 46's (D167-D193) were provided from the outset with a one-piece four-character headcode display panel.

The nose panel of Class 45 and 46 locomotives is a one-piece and interchangeable item. Spare panels were made after production of the locomotives themselves and were all of the final version, accommodating the single piece four-character code display. This panel has subsequently been fitted to all Class 46s; thus either type of divided display has come to provide a ready indication of a member of Class 45. Among Class 45's retrofitted with the single-piece headcode panels are Nos. 11, 21, 60, 82, 85, 90, 91 and 119. An oddity is No. 105, so fitted at one end and retaining at the other the original separate headcode boxes spaced widely apart; No. 71 is the same.

On the side of Class 45s and 46s, some locomotives have small extra ventilators some of which have subsequently been blanked off. Some side-mounted steps for access to the roof have also been covered over. A small triangular grille commonly, but not invariably, denotes a Class 46, this grille being located on the locomotive side beneath the long horizontal ventilator. Class 45's having the small triangular vent in use are the exception, examples being Nos. 11, 15, 18, 21, 60, 61 and 85.

LEADING PARTICULARS OF B.R. CLASS 44, 45 & 46 'PEAK' LOCOMOTIVES

	Class 44	Class 45	Class 46
Built by British Railways at	Derby	Derby and Crewe	Derby and Crewe
Power equipment: Diesel engine	12LDA28-A	12LDA28-B	12LDA28-B
	Sulzer Bros. Ltd.	Sulzer Bros. Ltd.	Sulzer Bros. Ltd.
Electrical equipment	Crompton Parkinson Ltd	Crompton Parkinson Ltd	Brush Elec. Eng. Co. Ltd
Wheel arrangement	I Co-Co I	I Co-Co I	I Co-Co I
Weight in working order	138 tons (with steam boiler)	136 tons	138 tons
Maximum axle load	19 tons	18 tons 16 cwt	19 tons
Adhesion weight	114 tons	112 tons 16 cwt	114 tons
Top speed	90 mph	90 mph	90 mph
Maximum tractive effort	70,000 lb	55,000 lb	55,000 lb
Continuous rated tractive	41,000 lb	30,000 lb	30,000 lb
effort at wheel rims	@ 16.5 mph	@ 25 mph	@ 25 mph
Engine hp (U.I.C.)	2,300 hp @ 750 rpm	2,500 hp @ 750 rpm	2,500 hp @ 750 rpm
Length over buffers	67 ft 11 in	67 ft 11 in	67 ft 11 in
Overall width	8 ft 10½ in	8 ft 10⅝ in	8 ft 10⅝ in
Height from rail to roof	12 ft 10½ in	12 ft 10½ in	12 ft 10½ in
Wheel diameter-driving	3 ft 9 in	3 ft 9 in	3 ft 9 in
Bogie pivot centres	32 ft 8 in	32 ft 8 in	32 ft 8 in
Bogie wheelbase-total	21 ft 6 in	21 ft 6 in	21 ft 6 in
Total wheelbase	59 ft 8 in	59 ft 8 in	59 ft 8 in
Minimum Radius Curve (locomotive alone)	330 ft (5 chains)	330 ft (5 chains)	330 ft (5 chains)
Fuel capacity – engine and boiler	840 gal	840 gal	840 gal
Boiler water capacity	1,340 gal	1,350 gal	1,350 gal
Main Generator Type	GC426-A1	GC426-A1	TG160-60
Traction motors (6)	C171-B1	C172-A1	TM73-68 Mk III

ORIGINAL ALLOCATIONS

D1-10 IB Camden from construction until 1962. Train heating removed then reallocation to Toton for freight-working only.

D11-31 55H Neville Hill.

D32-42 82A Bristol.

D43-137 17A Derby, or 14A Cricklewood. 93 Allocated direct to 82A for crew training in 1961 prior to introduction of 32-42.

D138-165 17A Derby.
138-154 crew training in 1962. [for example: 142 Leicester, 143 Saltley. 153 Haymarket. 154 Finsbury Park.] Before allocation to 17A.

D166-193 52A Gateshead.
155-165 tested from Derby on 1T48 working to Corby in 1962 then released to traffic; 166-193 tested on IT48 or run direct to Gateshead for traffic.

◄ *No. 46 005 (originally D142) at Penzance, July 1974. Although not endowed with the power, lightness or acceleration of Class 52 'Westerns', the Class 45 and 46 have proved highly competent performers on this arduous main line with its demanding schedules.* [*M. Wilson*]

ORIGINAL WORKINGS

44 (a) London-Manchester (Western Lines) when at 1B
(b) 16A only. Toton men trained on locos so work out and home from Toton to Sheffield, York, Washwood Heath, Wellingborough and Corby.

45 55H Midland Line Services from Leeds to Edinburgh and Glasgow, Leeds - Sheffield - Derby - Birmingham - Bristol and Leeds to St. Pancras. This covers both passenger and fast freight/parcels.
17A/14A Midland Main Line Services. Midland freight traffic, North East to South West Services (Later supplemented by Class 46 on reallocation).

46 17A North East-South West Services and Midland Line Services (gradually the 46 replaced 45 on NE-SW workings)
52A NE-SW Services, Liverpool-Newcastle workings and East Coast Services.

Later 46/47 became interchangeable and 52A class 46s were reallocated to 55A and 82A in exchange for Class 47 or 37. 17A Class 46s, were transferred to 82A in exchange for D32-42.

CLASS 45 RE-NUMBERING SCHEME

Steam-Heating Boiler only.

New No.	Old No.				
45 001	13	45 023	54	45 049	71
45 002	29	45 024	17	45 050	72
45 003	133	45 025	19	45 051	74
45 004	77	45 026	21	45 052	75
45 005	79	45 027	24	45 053	76
45 006	89	45 028	27	45 054	95
45 007	119	45 029	30	45 055	84
45 008	90	45 030	31	45 056	91
45 009	37	45 031	36	45 057	93
45 010	112	45 032	38	45 058	97
45 011	12	45 033	39	45 059	98
45 012	108	45 034	42	45 060	100
45 013	20	45 035	44	45 061	101
45 014	137	45 036	45	45 062	103
45 015	14	45 037	46	45 063	104
45 016	16	45 038	48	45 064	105
45 017	23	45 039	49	45 065	110
45 018	15	45 040	50	45 066	114
45 019	33	45 041	53	45 067	115
45 020	26	45 042	57	45 068	118
45 021	25	45 043	58	45 069	121
45 022	60	45 044	63	45 070	122
		45 045	64	45 071	125
		45 046	68	45 072	127
		45 047	69	45 073	129
		45 048	70	45 074	131

45 075	132	45 118	67
45 076	134	45 119	34
45 077	136	45 120	107
		45 121	18
		45 122	11
		45 123	52
		45 124	28
		45 125	123

Electric Train Heating only:

45 101	96	45 126	32
45 102	51	45 127	87
45 103	116	45 128	113
45 104	59	45 129	111
45 105	86	45 130	117
45 106	106	45 131	124
45 107	43	45 132	22
45 108	120	45 133	40
45 109	85	45 134	126
45 110	73	45 135	99
45 111	65	45 136	88
45 112	61	45 137	56
45 113	80	45 138	92
45 114	94	45 139	109
45 115	81	45 140	102
45 116	47	45 141	82
45 117	35	45 142	83
		45 143	62

45 144	55
45 145	128
45 146	66
45 147	41
45 148	130
45 149	135
45 150	78

NAMING OF LOCOMOTIVES OF CLASSES 44/45/46
(Original BR Numbers in brackets)

Number	Name	(BR No.)	Name carried earlier by		
44 001	Scafell Pike	(D1)			
44 002	Helvellyn	(D2)			
44 003	Skiddaw	(D3)			
44 004	Great Gable	(D4)			
44 005	Cross Fell	(D5)			
44 006	Whernside	(D6)			
44 007	Ingleborough	(D7)			
44 008	Penyghent	(D8)			
44 009	Snowdon	(D9)			
44 010	Tryfan	(D10)			
45 004	Royal Irish Fusilier	(D77)	Royal Scot	6123	
45 006	Honourable Artillery Company	(D89)	Royal Scot	6144	
45 104	The Royal Warwickshire Fusilier	(D59)	Royal Scot	6131	(Note 7)
45 111	Grenadier Guardsman	(D65)	Royal Scot	6110	
45 112	Royal Army Ordnance Corps	(D61)	Patriot	5505	
45 118	The Royal Artilleryman	(D67)	Royal Scot	6157	
45 039	The Manchester Regiment	(D49)	Royal Scot	6148	
45 040	King's Shropshire Light Infantry	(D50)	—	—	
45 123	The Lancashire Fusilier	(D52)	Royal Scot	6119	(Note 6)
45 041	Royal Tank Regiment	(D53)	Patriot	5507	(Note 4)
45 023	The Royal Pioneer Corps	(D54)	—		
45 144	Royal Signals	(D55)	Patriot	5504	
45 137	The Bedfordshire and Hertfordshire Regiment (TA)	(D56)	Patriot	5516	
45 043	The King's Own Royal Border Regiment	(D58)	Royal Scot	6136	(Note 5)
45 022	Lytham St Annes	(D60)	Patriot	5548	
45 143	5th Royal Inniskilling Dragoon Guards	(D62)	—		
45 044	Royal Inniskilling Fusilier	(D63)	Royal Scot	6120	
45	Coldstream Guardsman	(D64)	Royal Scot	6114	
45 046	Royal Fusilier	(D68)	Royal Scot	6111	
45 048	The Royal Marines	(D70)	—		
45 049	The Staffordshire Regiment (Prince of Wales' Own)	(D71)	Royal Scot / Royal Scot	6141 / 6143	(Note 1) / (Note 2)
45 055	Royal Corps of Transport	(D84)	Royal Scot	6126	(Note 3)
45 059	Royal Engineer	(D98)	Royal Scot	6109	
45 135	3rd Carabinier	(D99)	Royal Scot	6125	
45 060	Sherwood Forester	(D100)	Royal Scot	6112	(Note 8)
45 014	The Cheshire Regiment	(D137)	Royal Scot	6134	
46 026	Leicestershire and Derbyshire Yeomanry	(D163)	—		

Note 1: LMS No 6141 *The North Staffs Regiment*
Note 2: LMS No 6143 *The South Staffs Regiment*
Note 3: LMS No 6126 *Royal Army Service Corps*
Note 4: LMS No 5507 *Royal Tank Corps*
Note 5: LMS No 6136 *The Border Regiment*

Note 6: LMS No 6119 *Lancashire Fusilier*
Note 7: LMS No 6131 *The Royal Warwickshire Regiment*
Note 8: D100 was the first Class 45 to be named
Note 9: D163 was the only named Class 46

Class 45 D62, magnificently named 5th Royal Iniskilling Dragoon Guards, *at Exeter St. Davids with a Liverpool-Penzance train, July 1971. 'Peaks' have long been the mainstay of principal services from the Midlands and North into Devon and Cornwall.*

[H. L. Ford]